Driven to the Edge

The Untold Story of Hunt vs. Lauda

Etienne Psaila

Driven to the Edge: The Untold Story of Hunt vs. Lauda

Copyright © 2024 by Etienne Psaila. All rights reserved.

First Edition: **October 2024**

No part of this publication may be reproduced, distributed, or transmitted in any form or by any means, including photocopying, recording, or other electronic or mechanical methods, without the prior written permission of the publisher, except in the case of brief quotations embodied in critical reviews and certain other non-commercial uses permitted by copyright law.

ISBN: 978-1-923361-21-8

Table of Contents

Chapter 1: "Setting the Stage – Two Men, Two Worlds"

Chapter 2: "The Rise to Fame – Paths Converge on the Grand Stage"

Chapter 3: "Opposites in Overdrive – Hunt vs. Lauda's Racing Philosophies"

Chapter 4: "The 1976 Season Begins – Tensions Ignite"

Chapter 5: "High-Stakes on Every Corner – The Iconic Races"

Chapter 6: "The Tragedy at Nürburgring – Lauda's Defining Moment"

Chapter 7: "The Fighter Returns – Lauda's Remarkable Comeback"

Chapter 8: "Fuji Showdown – The Battle in the Rain"

Chapter 9: "Beyond the Podium – The Legacy of Hunt vs. Lauda"

Chapter 10: "Legends Forged in Fire – The Influence of a Rivalry"

Chapter 1: Setting the Stage – Two Men, Two Worlds

In the world of Formula 1, where grit and tenacity reign, James Hunt and Niki Lauda emerged as two of the most compelling figures. They were bound by their shared ambition to reach the pinnacle of motorsport, yet their paths, personalities, and driving philosophies couldn't have been more different. This opening chapter introduces the early lives of Hunt and Lauda, contrasting their backgrounds and the unique qualities that would set the stage for one of the greatest rivalries in the sport.

James Hunt: The Charismatic Playboy

James Simon Wallis Hunt was born in 1947 in Surrey, England, into an upper-middle-class family. His father, a stockbroker, held aspirations for his son to take on a respectable career, perhaps in finance or medicine. But even from a young age, Hunt's spirited and rebellious nature hinted at a different path. At school, he was known for his quick wit and equally quick temper, traits that would follow him throughout his career.

Hunt's fascination with speed and thrill came somewhat by chance. As a teenager, he dabbled in various sports, excelling in tennis, squash, and skiing. It wasn't until he was 18 that he attended a race at Silverstone and found himself mesmerized by the world of motorsport. From that moment on, his life took a decisive turn. The world of Formula 1, with its high stakes and visceral thrill, offered exactly the sort of freedom Hunt craved.

However, his rise was neither swift nor smooth. In the competitive and financially demanding world of British racing, Hunt's early career was marked by struggles to find sponsorship and financial backing. But his larger-than-life personality, coupled with undeniable raw talent, soon caught the attention of the paddock. Known for his unpredictable driving style, Hunt was often branded as reckless, yet he had a unique ability to maneuver his car with an instinctive grace that defied logic.

Hunt became as famous off the track as he was on it. Known for his playboy lifestyle, he was unapologetically hedonistic, often spotted in the company of beautiful women, partying until the early hours, and living by his own rules. This carefree, devil-may-care approach endeared him to fans and made him a media sensation. But behind the relaxed exterior was an intense competitor whose hunger for victory drove him forward. For Hunt, racing was about the thrill and the freedom of the moment, about pushing himself—and his car—to the absolute limit.

Niki Lauda: The Disciplined Perfectionist

Andreas Nikolaus "Niki" Lauda was born in 1949 in Vienna, Austria, into a wealthy family of industrialists. His family's prosperity offered him a comfortable life, but Lauda found himself at odds with their ambitions for him. The Lauda family expected Niki to follow in their footsteps, studying business or engineering to maintain their legacy. But young Lauda had different ideas. Fascinated by cars from an early age, he was singularly determined to become a professional driver—a pursuit his family saw as both frivolous and dangerous.

Unlike Hunt, Lauda's approach to racing was meticulous, calculating, and disciplined. Every decision he made was a calculated step toward his ultimate goal of winning championships. His ascent in motorsport was methodical; he began with small races, applying his analytical mind to understand every aspect of driving and mechanics. His ability to communicate effectively with his engineers, providing precise feedback on car performance, made him a valuable asset to any team. But Lauda's family disapproved so strongly of his career choice that they cut him off financially, forcing him to take out loans to fund his career and buy his way into a Formula 2 seat.

While Hunt was known for his flamboyance, Lauda's reserved nature often masked the fire within. He was determined, even ruthless, in his pursuit of success. For Lauda, every move on and off the track was part of a larger strategy to win. In the cockpit, he was relentless, analyzing every corner, every lap, in pursuit of perfection. He believed in the power of discipline, a trait that sometimes made him appear cold and unapproachable to others in the paddock. But for Lauda, racing wasn't just about thrill—it was about control, precision, and the cold-blooded drive to be the best.

Early Career Parallels and Divergences

The world of Formula 1 is a melting pot of personalities, but few stood out as sharply as Hunt and Lauda. In the early years of their careers, they raced in parallel, each encountering their share of challenges. Hunt, with his carefree approach, took a more erratic path. He signed with smaller teams, scraping together enough sponsorship to keep himself on the grid. His risk-taking style often ended in disaster, earning him the nickname "Hunt the Shunt," yet his perseverance and daring moves kept fans captivated.

Lauda's approach was diametrically opposite. After his family cut him off, he borrowed heavily to secure a position at March Engineering. Although he began with a disadvantage, his consistent performance and methodical feedback soon earned him a reputation as a promising driver. His technical feedback and understanding of the car's limits impressed his engineers, paving the way for a move to BRM, where he further honed his craft.

Their contrasting personalities began to create a subtle tension. Hunt was all about natural talent and unrestrained thrill, while Lauda treated driving like a science, dissecting every component and never leaving room for error. By the time they reached Formula 1, they had each carved out their distinctive reputations. Hunt, the brash and charming maverick, versus Lauda, the disciplined perfectionist who left nothing to chance. The motorsport world, sensing the potential for a showdown between these two, awaited their inevitable clash.

The Unlikely Rivals

As they rose through the ranks of Formula 1, Hunt and Lauda's rivalry simmered beneath the surface. They shared a mutual respect for each other's abilities, but their clashing approaches often put them at odds. Hunt's natural charisma and disregard for convention contrasted sharply with Lauda's relentless discipline, sparking a rivalry that was as much about values as it was about driving skill. Hunt believed in the freedom of speed, while Lauda saw it as a matter of calculated control—a philosophical difference that would fuel their competition.

While Hunt garnered a legion of fans who adored his rebel persona, Lauda cultivated a loyal following for his unyielding work ethic and dedication to his craft. Their paths finally converged in the 1976 season, but before that year, each driver had already laid the foundation of his legacy. The stage was set for one of the most intense rivalries in Formula 1 history—a battle not only of skill but of opposing ideologies that would soon captivate the world.

Chapter 2: The Rise to Fame – Paths Converge on the Grand Stage

The journey to Formula 1 stardom is never easy, but for James Hunt and Niki Lauda, it was a story of defying odds, leveraging raw talent, and embodying the distinct approaches that would ultimately define their rivalry. As they entered Formula 1, their paths began to converge, setting them on a collision course that would lead to one of the most legendary seasons in racing history.

James Hunt: The Maverick's Entrance

James Hunt's entry into Formula 1 was as unorthodox as his personality. By the early 1970s, he had developed a reputation as a daring, sometimes reckless driver, willing to push limits that others would have feared. Hunt's early racing career was marked by struggles with financing and consistency; he started in the British Formula 3 series, where his tendency to push too hard often ended with his car in a cloud of dust. His aggressive style earned him the nickname "Hunt the Shunt"—a reflection of his many crashes—but it also marked him as a driver to watch. Despite his crashes, Hunt had undeniable charisma and driving talent, qualities that began attracting sponsors and supporters who were willing to take a chance on him.

Hunt's big break came with Hesketh Racing, a small private team led by the eccentric Lord Alexander Hesketh. Unlike most Formula 1 teams, which operated with strict budgets and a focus on performance, Hesketh Racing exuded an air of irreverence and extravagance, which suited Hunt perfectly. The team's relaxed attitude allowed Hunt to thrive, and he quickly began to show his potential. Hesketh Racing lacked the resources of the big teams, but Hunt's talent behind the wheel kept them competitive. His first taste of real success came in 1975 when he scored his first Formula 1 victory at the Dutch Grand Prix, a win that was as much a testament to Hunt's raw skill as it was to his determination to prove himself.

With his first win, Hunt's reputation exploded. The press adored his freewheeling lifestyle and flamboyant personality, and fans loved his aggressive driving and candid attitude. Hunt represented a break from the polished, corporate image of Formula 1 drivers, embracing a rebellious, rock-and-roll persona that was rare in the paddock. His popularity grew, but with it came pressure to prove that he wasn't just a one-time winner. The racing world was beginning to realize that behind Hunt's easygoing facade was a fierce competitor, hungry for the thrill of victory.

Niki Lauda: The Calculated Ascent

Niki Lauda's path to Formula 1 was a study in contrasts. While Hunt rode on charisma and natural talent, Lauda's rise was grounded in hard work, determination, and an unshakable belief in discipline. After financing his early racing career with loans and breaking away from his family's expectations, Lauda methodically worked his way up through the racing ranks, starting with Formula 2 and smaller Formula 1 teams. His approach to driving was deliberate, almost scientific, and his feedback to engineers was detailed and precise—qualities that quickly gained the attention of those in the racing world.

Lauda's big break came in 1974, when Enzo Ferrari, impressed by his technical feedback and consistent results, offered him a contract with Scuderia Ferrari. Ferrari was a team steeped in legacy, and Lauda was thrilled to join its ranks, seeing it as the ultimate validation of his efforts. Unlike Hesketh Racing, Ferrari was highly disciplined, with a rigorous focus on performance and precision—an environment that suited Lauda perfectly. Lauda threw himself into the role, working closely with Ferrari's engineers to refine the car's performance and ensuring that every detail was optimized for success.

In 1974, Lauda proved his worth, taking his first victory at the Spanish Grand Prix. His triumph wasn't just a victory for Ferrari but a clear statement of his driving prowess and technical understanding. Lauda quickly established himself as a serious contender, someone who, despite his youth, possessed the maturity and focus of a seasoned driver. His approach was in stark contrast to Hunt's—where Hunt was passionate and impulsive, Lauda was cool and calculating, determined to minimize risk at every turn. His methodical style paid off, as he secured a series of podium finishes, earning him the respect of both his team and competitors.

Earning Their Reputations

As Hunt and Lauda gained traction in Formula 1, their reputations crystallized in ways that mirrored their personalities. Hunt's racing style was fast, aggressive, and at times reckless, yet undeniably thrilling. Fans adored him for his all-or-nothing attitude, his willingness to take risks that left others gasping. Off the track, his lifestyle made him a media darling; he partied hard, entertained fans with his brash humor, and lived life as if each race could be his last. Hunt's natural charm made him a fan favorite, but his unconventional methods kept critics on edge, questioning his long-term potential in the sport.

Lauda, on the other hand, was regarded as a perfectionist, someone who left nothing to chance. His intense focus on understanding every technical aspect of his car led to a string of consistently strong performances, and his disciplined approach earned him admiration and respect. Where Hunt was loose and unpredictable, Lauda was precise and measured. This calculated style made him less sensational in the media but highly respected within the paddock. Lauda represented the ideal of a true professional, someone who approached every race with a clear plan and exacting standards. This image of Lauda as the ultimate professional began to resonate with fans who appreciated the discipline he brought to a sport often associated with high-stakes glamour and drama.

The Convergence

By the mid-1970s, both Hunt and Lauda had firmly established themselves in Formula 1, albeit through different paths and styles. While Hunt was known for his unpredictability and crowd-pleasing antics, Lauda was seen as the reliable tactician who could always be counted on to deliver consistent results. Their rivalry had yet to reach its peak, but both men were now stars of the sport, each one embodying an archetype that drew fans from around the world.

The stage was set for the 1976 season, a year that would prove to be a turning point in both their careers. Lauda was coming off a highly successful 1975 season in which he won his first World Championship with Ferrari, solidifying his place as the sport's top driver. For Lauda, the 1976 season represented an opportunity to defend his title and continue his legacy as one of the most disciplined and skilled drivers in Formula 1.

Meanwhile, Hunt was entering the season with a fresh contract at McLaren, a team with the resources and experience to support a championship contender. Hesketh Racing, while thrilling, could no longer keep up with the top teams, and Hunt's move to McLaren was a critical step in his career. With McLaren's support, he now had a car capable of challenging Lauda's Ferrari and the backing of a team that shared his ambition for glory.

Their paths, once parallel but separate, were now set to collide. Lauda, the reigning champion, saw the season as a means to reinforce his dominance, while Hunt, the maverick challenger, viewed it as the perfect opportunity to break into the ranks of champions. Formula 1 fans around the world sensed the impending showdown between two drivers who couldn't be more different—one who represented discipline and calculation, the other passion and spontaneity.

The 1976 Season Awaits

As the 1976 season loomed, the anticipation was palpable. Both Hunt and Lauda entered the season with something to prove. Lauda was defending his title, and his disciplined approach made him the favorite. But Hunt, now with a competitive car, was eager to make his mark. The two drivers had evolved in their own distinct ways, but their rivalry was about to take on new dimensions, transforming from a contest of skills to a battle of ideologies. It was the meticulous versus the audacious, the structured versus the spontaneous, in a high-stakes race for glory.

This season would challenge them in ways they could never anticipate, both on and off the track. It would test their physical limits, their emotional resilience, and the very philosophies they lived by. For the world of Formula 1, the rivalry was about to become one of the greatest narratives in the sport's history—a story of two men, their competing visions, and the battle to see which one would emerge as the true champion.

In the following chapters, we'll journey through this incredible season, exploring the races, victories, crashes, and comebacks that turned Hunt and Lauda from competitors into legends. The rise to fame had set the stage, but the 1976 season would be the crucible in which their legacies were forged.

Chapter 3: Opposites in Overdrive – Hunt vs. Lauda's Racing Philosophies

On the track, where every split-second decision could spell the difference between victory and disaster, James Hunt and Niki Lauda adopted starkly contrasting philosophies. The thrill of racing, the push toward perfection, and the desire to become world champion were ambitions they shared, but the ways they approached these goals couldn't have been more different. Hunt's style was fiery and impulsive, while Lauda's was calm and calculated—a juxtaposition that would not only define their rivalry but also create an unforgettable dynamic that captivated the world.

Hunt's Racing Philosophy: All or Nothing

James Hunt drove with a fearlessness that bordered on reckless abandon, embracing a style that was both unpredictable and electrifying. For Hunt, racing was a thrill to be seized, an experience to be savored in the moment without worrying too much about the consequences. In his mind, the best way to race was with unfiltered passion, pushing himself and his car to the limit, sometimes beyond. He believed that victory came from taking risks, and in his world, there was no glory in half measures.

Hunt's approach was to drive with instinct and a natural feel for the car, something he relied on more than technical calculations or meticulous preparation. He didn't spend hours analyzing data or optimizing his technique. Instead, he trusted his gut, and for him, this was the purest form of racing. In his early career, this instinctive style often led to spectacular crashes as he flirted with the edges of control, but it also led to moments of sheer brilliance when his natural talent shone through. To Hunt, racing was as much about the thrill as it was about winning. He wanted to feel the rush, to live dangerously, and to enjoy every second of it.

Hunt's all-or-nothing approach to racing mirrored his approach to life off the track. Known for his hedonistic lifestyle, he indulged in fast living, hard partying, and casual relationships, becoming a symbol of the rock-and-roll spirit in Formula 1. His race-week rituals were often unconventional—nights filled with champagne, cigarettes, and laughter. Hunt had no qualms about showing up to the track slightly hungover or wearing a "sex, breakfast of champions" patch on his overalls. His carefree attitude off the track matched his racing style, reinforcing his image as a playboy with a penchant for thrill-seeking.

This approach, however, came with its own set of challenges. Hunt's reliance on instinct and disregard for consistency meant that he often struggled to maintain focus or control, particularly when things didn't go his way. His emotions ran high, and he was known for outbursts and moments of frustration, especially when faced with technical issues. Hunt's fans loved him for his spontaneity and charisma, but his peers often doubted his long-term potential, questioning whether he had the discipline to reach the top.

Lauda's Racing Philosophy: Calculated Precision

In contrast, Niki Lauda's approach to racing was grounded in careful analysis, preparation, and control. Where Hunt thrived on the thrill of spontaneity, Lauda found power in order and precision. He believed that winning wasn't just a matter of talent but also of preparation, strategy, and an unwavering commitment to improvement. His race preparation was methodical: he studied each track, memorized every corner, and worked closely with his team of engineers to optimize the car's performance. For Lauda, no detail was too small to overlook, and he left nothing to chance.

Lauda approached each race like a puzzle to be solved, breaking down every aspect of his car and understanding it from a technical perspective. He spent hours with his team, fine-tuning the car to his exact specifications, and gave feedback that was so precise it was almost surgical. His feedback helped Ferrari make significant improvements, and it wasn't long before his skill for diagnosing mechanical issues became legendary. Lauda's driving style was also calculated; he knew how to push the car's limits without endangering himself or the equipment, preferring a consistent performance over flashier, risky maneuvers. It was an approach built on minimizing risks and maximizing efficiency.

Off the track, Lauda's lifestyle was a reflection of this philosophy. He was known for his quiet, disciplined life, largely free from the glamour and excess that surrounded Formula 1. Unlike Hunt, Lauda kept to a strict routine, focused on maintaining peak physical and mental condition. His approach to life was no-nonsense, just as it was in racing. He avoided distractions and steered clear of the party scene, preferring solitude or time spent with his engineers, deep in discussion about mechanics or strategy. This level-headedness, though sometimes mistaken for coldness, was Lauda's way of staying grounded. He respected Formula 1 not as a playground for thrills but as a professional sport that demanded complete focus and respect for the machinery.

Lauda's approach did have its drawbacks, though. His intense focus on consistency and aversion to risk meant he sometimes lacked the audacious flair of a driver like Hunt. Critics would occasionally question whether Lauda was "too safe" to be a champion, arguing that his methodical style lacked the aggressiveness that fans craved. However, Lauda's results on the track silenced most critics; his method worked, allowing him to perform consistently at a high level. He may not have been the showman Hunt was, but Lauda's precision made him one of the most formidable drivers in Formula 1.

A Philosophy Clash on the Track

Their contrasting philosophies created a dynamic that was palpable every time they lined up on the grid. Hunt's all-or-nothing style meant that he often took more risks on the track, darting into corners, making bold overtakes, and occasionally pushing the boundaries of control. To him, every position gained was worth the gamble. For Lauda, however, racing was a disciplined exercise in maintaining the optimal line and pace. He took fewer risks, always weighing the long-term consequences of his actions. While Hunt would charge forward with aggressive overtakes, Lauda would calculate the best moment to strike, often choosing strategy over spectacle.

During races, this difference in style was obvious not only to the spectators but also to other drivers, who would occasionally take sides in the rivalry, aligning themselves with Hunt's fearless drive or Lauda's controlled mastery. When the two went wheel-to-wheel, it was a spectacle: Hunt, willing to risk everything for the immediate gain, versus Lauda, carefully waiting for the right moment to capitalize on a mistake. Each driver's presence on the track forced the other to refine his approach, adding an extra layer of tension and excitement to every race.

Their approaches also affected how they handled adversity. When faced with mechanical issues or poor qualifying results, Hunt would react with frustration, occasionally letting his emotions get the best of him. Lauda, on the other hand, remained cool-headed, using setbacks as opportunities to analyze and learn. He would methodically assess what went wrong and work with his team to prevent it from happening again, an approach that eventually won him the respect of his competitors.

The Tension Builds

These two distinct racing philosophies and lifestyles created a tension that was felt both on and off the track. Hunt's flamboyant, unpredictable nature stood in stark contrast to Lauda's calm professionalism. They were civil but never truly friends; each respected the other's talent but disagreed fundamentally on what it meant to be a racing driver. Hunt often joked about Lauda's reserved demeanor, and Lauda, in turn, privately questioned Hunt's commitment to racing as a serious sport. This undercurrent of mutual critique added fuel to their rivalry, turning each race into a psychological as well as physical battle.

Fans were also split, with some drawn to Hunt's rebellious image and others admiring Lauda's disciplined focus. The media, sensing the appeal of this rivalry, played up their differences, portraying Hunt as the charismatic hero and Lauda as the calculating tactician. Each driver, in his own way, represented an ideal that resonated deeply with audiences. Hunt was the embodiment of living in the moment, seizing the thrill of life, while Lauda stood for precision, order, and dedication.

The Road to the 1976 Showdown

As the 1976 season approached, the intensity of their rivalry reached new heights. Lauda, coming off a World Championship, was seen as the man to beat, his disciplined approach placing him at the top of the Formula 1 world. Hunt, now with McLaren, finally had a car that could match Ferrari's power. For the first time, he had the resources to challenge Lauda seriously, and he was determined to make the most of it. Each driver entered the season with everything to gain—and everything to prove.

This year would test not just their driving skills but the very philosophies they lived by. Lauda's disciplined precision would be pitted against Hunt's raw passion, and every race would be a test of who could better withstand the pressure and bring their unique style to victory. As the season opener loomed, Formula 1 fans braced themselves for a battle unlike any other—a showdown between two men who embodied the very soul of the sport yet couldn't have been more different in their approaches.

The stage was set for the ultimate test: a season that would demand every ounce of their strength, determination, and skill. It would push them to the brink, shaping not only their careers but also their lives in ways they could never have imagined.

Chapter 4: The 1976 Season Begins – Tensions Ignite

The 1976 Formula 1 season began with a sense of palpable anticipation. For fans, it was a season expected to showcase a fierce battle between two of the sport's most captivating personalities. Niki Lauda, the reigning World Champion, had everything to lose, while James Hunt, with his newfound position at McLaren, had everything to gain. Their contrasting approaches, which had already caused sparks in previous seasons, were now set to ignite in an unprecedented showdown. As the season opener drew near, the atmosphere within both McLaren and Ferrari reflected the high stakes and brewing tension.

A Tale of Two Teams: McLaren and Ferrari

For McLaren, Hunt's arrival marked a fresh start. The team, led by the legendary Teddy Mayer, had a competitive car with the McLaren M23—a proven design capable of going toe-to-toe with Ferrari. While Hunt's flamboyant style was known to cause some concern among team members, Mayer saw his star driver's potential and believed Hunt had the raw skill to challenge Lauda. McLaren, accustomed to drivers like the cool-headed Emerson Fittipaldi, found themselves adjusting to Hunt's less conventional approach. The team's focus was clear: they had invested in Hunt with the belief that he could bring McLaren back to the top, and they were willing to adapt to his eccentricities if it meant securing victory.

Meanwhile, at Ferrari, Lauda entered the season with high expectations. Enzo Ferrari and the Scuderia Ferrari team held their drivers to an extraordinarily high standard, and Lauda's 1975 championship season had reinforced his reputation as the ultimate professional. The team's infrastructure and resources were unmatched, and Lauda, with his technical expertise and methodical style, seemed poised for another championship run. Yet, despite his dominance, Lauda sensed that Hunt's arrival at McLaren posed a genuine threat. The Ferrari team's atmosphere was one of intense discipline, and Lauda, ever the strategist, approached the season with his trademark precision, ready to defend his title against the unpredictable British contender.

The Season Opener: Brazil – A Slow Burn

The season opened at the Brazilian Grand Prix in January 1976. The hot, humid conditions tested drivers' endurance, but Lauda's disciplined preparation allowed him to perform consistently, securing a respectable third-place finish. Hunt, still finding his footing with McLaren, had an uneven performance and struggled with car issues that kept him from finishing high in the standings. For Hunt, it was a reminder of the challenges he would face, not just from Lauda but also from the grueling demands of the season.

Despite the rocky start, Hunt's potential was evident. The McLaren team saw glimpses of his raw skill, and fans were drawn to his aggressive style. Lauda, while off to a smooth start, remained vigilant. He understood that Hunt's competitive fire would grow as he adjusted to the car, and Lauda knew he needed to keep his lead while he could. Their rivalry simmered quietly, awaiting its chance to ignite.

The Spanish Grand Prix: First Sparks Fly

By the time the season reached Spain, Hunt was beginning to hit his stride with McLaren. He was driving with increasing confidence, and his connection with the car was evident. The Spanish Grand Prix at Jarama marked the first serious clash between Hunt and Lauda, where the tension between their driving styles became a focal point.

In Spain, Hunt put on a masterclass, showcasing his aggressive, risk-taking style as he fought his way to the front. His ability to overtake, often in the most improbable situations, kept fans on edge, and by the end of the race, he crossed the line first. However, the celebration was short-lived. Technical inspections revealed that Hunt's McLaren was slightly too wide, leading to his disqualification—a decision that handed the win to Lauda.

The disqualification hit Hunt hard, igniting a sense of injustice within him that added fuel to the rivalry. McLaren appealed the decision, launching a lengthy protest that would ultimately result in Hunt's victory being reinstated months later. But in the moment, Hunt's frustration was palpable. Lauda, on the other hand, maintained his composure, expressing a professional detachment and accepting the victory with a calm, almost dismissive attitude. For Lauda, it was simply another step in his campaign to defend his title, but for Hunt, it was a spark that stoked the fires of his determination.

Monaco: Precision vs. Passion

The Monaco Grand Prix, a jewel in the Formula 1 calendar, provided a perfect setting to showcase the stylistic clash between Hunt and Lauda. Monaco's narrow, winding streets required absolute precision, leaving no margin for error. Lauda, known for his disciplined and measured driving, thrived on the challenging circuit. He executed each turn with calculated precision, ultimately winning the race in dominant fashion.

Hunt, by contrast, struggled in Monaco. The tight circuit restricted his aggressive style, forcing him to rein in his natural impulses. It was a frustrating race for him, where Lauda's mastery of control underscored the advantage of his calculated approach. The contrast between their performances highlighted the growing tension in their rivalry: while Hunt's daring style had its advantages, Lauda's disciplined approach was proving more reliable.

In the paddock, the rivalry between the two drivers was becoming the main story. Journalists and fans alike began to follow their every move, eager to see who would come out on top in each race. Hunt's fiery temperament contrasted sharply with Lauda's calm resolve, and their interactions—while respectful—were laced with subtle jabs and competitive tension. Each man was driven not only by the desire to win but also by the desire to prove that his way was superior.

Belgium: Hunt's Redemption

The Belgian Grand Prix at Zolder was a turning point for Hunt. Determined to close the gap to Lauda, he entered the race with a renewed focus. From the moment he arrived at the circuit, Hunt seemed different—more focused, more determined. It was as if the frustrations of the season's early races had crystallized into a fierce resolve.

Hunt's performance in Belgium was a revelation. He drove with a controlled aggression, balancing his natural instincts with a newfound discipline. This approach paid off, as he dominated the race and took his first official win of the season. The victory was a statement—not just to Lauda but to the entire paddock—that Hunt was a serious contender.

Lauda, who finished third, took note of Hunt's improved performance. For the first time, he saw that Hunt was capable of controlling his instincts, a development that made him an even more formidable opponent. The victory in Belgium marked a shift in Hunt's mentality; he now saw himself as Lauda's equal, and he was determined to prove it in every race.

The Pressure Builds: Sweden and Beyond

As the season progressed to the Swedish Grand Prix, the competition between Hunt and Lauda grew increasingly intense. Lauda, with his characteristic consistency, continued to perform at a high level, securing podium finishes that kept him comfortably in the lead for the championship. Hunt, however, was closing the gap, driven by his passion and the desire to prove himself.

In Sweden, Hunt and Lauda found themselves in close quarters once again, and the race was marked by a series of tense moments as the two battled for position. Lauda's calm control was on full display, but Hunt's relentless pressure kept him on edge. The race was a testament to the clash of their driving philosophies: Hunt, taking risks to gain every possible advantage, versus Lauda, maintaining his pace with unwavering focus.

Despite Lauda's lead in the championship standings, Hunt's consistent pressure was beginning to take a toll. The rivalry had evolved into a psychological battle, with each driver pushing the other to new extremes. Every podium finish, every overtake, every point earned became a message. For Lauda, it was a reminder of his superiority as a disciplined driver; for Hunt, it was a statement of his determination to win at any cost.

A Rivalry in Full Force

By the mid-season point, the 1976 championship had become a two-man race between Hunt and Lauda. Their rivalry, already intense, had reached new heights, with each driver embodying a distinct approach to racing. The fans were drawn into the drama, taking sides and cheering for their chosen champion. Hunt's raw charisma and aggressive style made him a fan favorite, while Lauda's precision and professionalism earned him the respect of purists who admired his methodical approach.

For both men, the stakes were higher than ever. Hunt wanted to prove that his instinct-driven, passionate style was just as valid as Lauda's disciplined control, while Lauda aimed to defend his title and affirm the superiority of his calculated approach. The battles they fought on the track mirrored a deeper philosophical struggle, one that captivated the world and set the stage for the dramatic events to come.

The 1976 season was far from over, and as the rivalry intensified, so did the risks. In their pursuit of victory, both Hunt and Lauda would be tested in ways they could never have anticipated. The opening races had only been a prelude to a showdown that would push them to the brink, physically, mentally, and emotionally. As the season continued, Formula 1 would witness a rivalry for the ages, a battle between two men with everything to prove and nothing to lose.

Chapter 5: High-Stakes on Every Corner – The Iconic Races

As the 1976 season unfolded, the Formula 1 world watched in awe as Niki Lauda and James Hunt battled through some of the most intense and memorable races in the sport's history. Every corner, every overtake, and every lap was charged with tension as the two rivals pushed each other to their limits. For both men, these races weren't merely contests of skill but tests of character, willpower, and endurance. Their differences and fierce drive to outdo one another elevated each race into a spectacle, leaving fans on the edge of their seats and setting the stage for an unforgettable championship.

The British Grand Prix at Brands Hatch – A Controversial Showdown

The British Grand Prix at Brands Hatch was one of the most controversial and fiercely contested races of the season. For Hunt, it was a homecoming—a chance to prove himself in front of British fans who adored his rebellious charm. The atmosphere at Brands Hatch was electric as fans packed the stands, waving Union Jacks and cheering on their hero. Hunt, who thrived on crowd energy, was determined to give them a performance to remember.

The race began with high drama. On the opening lap, a collision between Hunt, Lauda, and Regazzoni forced the race to be stopped, much to the frustration of drivers and fans alike. After a delay, the race was restarted, but not without controversy. Initially, Hunt was told he could not restart because he had returned to the pits before the red flag was raised—a violation of race protocol. The fans at Brands Hatch erupted in anger, chanting Hunt's name and demanding his reinstatement. Under intense pressure from the crowd and the McLaren team, the stewards ultimately allowed Hunt to restart the race.

Once back on track, Hunt seized the opportunity with fierce determination. He drove with an aggression that matched the passion of the fans, overtaking Lauda and several others in a spectacular display of skill. Hunt eventually took the checkered flag, sending the crowd into a frenzy. However, the victory was short-lived. Ferrari filed an official protest, arguing that Hunt's participation in the restart was against the rules. Weeks later, the appeal was upheld, and Hunt's victory was annulled, handing the win to Lauda.

The controversial ruling further fueled the rivalry. Hunt,

frustrated by what he saw as a politically motivated decision, vowed to push even harder in the remaining races. Lauda, though pleased with the outcome, remained focused, seeing the controversy as just another obstacle in his campaign to retain the championship. This race encapsulated the tension between the two men, showcasing Hunt's fiery passion and Lauda's calculated resilience.

The Dutch Grand Prix at Zandvoort – Hunt's Redemption

The Dutch Grand Prix at Zandvoort offered Hunt an opportunity to redeem himself and show that he could win without controversy. Zandvoort, with its challenging curves and sandy terrain, was a track that demanded both skill and finesse. Hunt arrived in the Netherlands with a renewed focus, determined to prove that his talent could outshine any disputes or disqualifications.

From the start, Hunt was in top form. He pushed his McLaren to its limits, navigating Zandvoort's tight corners and undulating curves with a blend of aggression and precision that mesmerized fans. Lauda, too, was strong, maintaining his characteristic consistency, but Hunt's speed was unmatched. He took the lead early on, fending off Lauda with daring overtakes and defending his position through a series of intense battles.

The race was a showcase of Hunt's tenacity. Every corner and straight became a testament to his all-or-nothing approach, and his performance at Zandvoort silenced many critics who had questioned his legitimacy as a championship contender. Hunt crossed the finish line victorious, proving that he could defeat Lauda fair and square. For Hunt, Zandvoort was more than a victory—it was a statement. He had shown the world that he was not just a playboy racer but a serious competitor who could go toe-to-toe with the reigning champion.

Lauda, though gracious in defeat, recognized that Hunt's win at Zandvoort posed a significant threat to his title defense. The rivalry had evolved beyond mere competition; it was now a battle for supremacy, with each race amplifying the stakes.

The French Grand Prix at Paul Ricard – Lauda's Answer

After Hunt's triumph in the Netherlands, the championship race tightened, and Lauda sought to reassert his dominance at the French Grand Prix at Paul Ricard. The French track, known for its long straights and challenging high-speed corners, was ideal for Lauda's calculated style. He approached the race with a renewed intensity, determined to remind Hunt and the world why he was the reigning champion.

Lauda's performance in France was clinical. From qualifying to the checkered flag, he displayed flawless control, setting a pace that no one, including Hunt, could match. Where Hunt often took risks, Lauda relied on his discipline, choosing his moments carefully and conserving his car's resources. His methodical approach paid off as he claimed victory with a significant lead, widening his points gap and restoring his confidence.

This race demonstrated Lauda's strategic prowess. While Hunt thrived in races that required split-second decisions and daring moves, Lauda excelled in conditions that rewarded patience and planning. His win in France was a reminder that the championship wasn't simply about raw speed; it was also about consistency, strategy, and focus. Hunt, though disappointed, couldn't deny Lauda's skill, and the French Grand Prix added another layer of respect to their rivalry.

The German Grand Prix at Nürburgring – A Defining Moment

The German Grand Prix at the Nürburgring, known as the "Green Hell," was one of the most treacherous circuits in Formula 1. Stretching over 14 miles with 154 turns, the Nürburgring demanded a level of bravery and skill that few could muster. For Hunt and Lauda, this race would become a defining moment—not only for the season but for their lives.

Lauda, who had voiced concerns about the track's safety, considered boycotting the race, but the event went on as scheduled. The weather on race day was unpredictable, with rain leaving parts of the track slick and treacherous. Hunt, undeterred by the conditions, prepared for the race with his characteristic courage. Lauda, though uneasy, decided to compete, determined to protect his championship lead.

Tragically, on the second lap, Lauda lost control of his Ferrari, crashing into a barrier. His car burst into flames, and Lauda was trapped inside, suffering severe burns and inhaling toxic fumes before being rescued by fellow drivers. The accident was catastrophic, leaving Lauda with life-threatening injuries that forced him out of racing indefinitely.

For Hunt, the crash was a sobering moment. The rivalry between them suddenly seemed secondary to Lauda's life and well-being. As Hunt continued through the race, his thoughts were with Lauda, and he later dedicated his win to his rival. Lauda's accident would haunt Hunt and the entire racing world, underscoring the dangers of Formula 1 and forever changing the nature of their rivalry.

The Italian Grand Prix at Monza – Lauda's Miraculous Comeback

Barely six weeks after his devastating accident, Lauda stunned the world by returning to racing at the Italian Grand Prix at Monza. Against the advice of doctors, who feared for his recovery, Lauda insisted on competing, determined to defend his title. His face was still visibly scarred, and he wore specially designed bandages to protect his healing skin, but his resolve was unshaken.

The sight of Lauda on the Monza grid was nothing short of miraculous. He qualified in the top four and drove with an intensity that defied belief, finishing fourth and securing crucial points in his championship battle with Hunt. Lauda's return at Monza was a testament to his indomitable spirit and his refusal to let anything stand in his way.

Hunt, who had won two races in Lauda's absence, was deeply moved by Lauda's comeback. He expressed admiration for his rival, recognizing that Lauda's resilience and courage transcended their rivalry. The Monza race reignited the championship battle, with Lauda now back in contention, and it added a new dimension to the Hunt-Lauda dynamic—one that went beyond competition and into mutual respect.

The Canadian and U.S. Grand Prix – The Battle Continues

As the season neared its climax, Hunt and Lauda fought fiercely at the Canadian and U.S. Grands Prix, trading wins and podium finishes. The intensity of their rivalry had reached a fever pitch, with each race feeling like a championship decider. Fans around the world were captivated, watching two drivers who, despite their differences, were evenly matched and driven by an unyielding desire to win.

At the Canadian Grand Prix, Hunt took the victory, closing the points gap and increasing the pressure on Lauda. The U.S. Grand Prix at Watkins Glen saw Lauda bounce back, finishing on the podium and maintaining his narrow lead. Each race added fuel to the fire, with both drivers pushing themselves and their cars to the absolute limit.

Setting the Stage for Fuji – A Championship Decider

By the time the championship reached its final race in Japan, Hunt and Lauda were separated by only a few points. The tension between them was at an all-time high, with both men knowing that the Fuji showdown would determine who would claim the 1976 title. Hunt, riding high on a string of recent victories, entered the race with renewed confidence, while Lauda, who had fought back from the brink of death, was determined to see his journey through to the end.

The stage was set for one of the most dramatic finales in Formula 1 history. Each race of the 1976 season had brought Hunt and Lauda closer to this moment, where their contrasting styles, fierce rivalry, and unyielding spirit would collide one last time. The world watched as two men who had given everything they had—physically, mentally, and emotionally—prepared for a final battle that would define their legacies.

The Fuji showdown would be a race for the ages, a culmination of a season marked by triumph, tragedy, and a rivalry that transcended the sport.

Chapter 6: The Tragedy at Nürburgring – Lauda's Defining Moment

The German Grand Prix at Nürburgring, famously nicknamed the "Green Hell," is one of the most challenging and unforgiving circuits in the world. Stretching over 14 miles with 154 perilous turns, it had claimed many lives and posed a serious threat to every driver who dared to race on it. For Niki Lauda, however, the Nürburgring was not just a race; it became the site of a tragedy that would change his life and the world of Formula 1 forever.

Concerns and Warnings

In the days leading up to the German Grand Prix on August 1, 1976, Lauda, along with other drivers, voiced his concerns about the safety of the Nürburgring. The track's vast length and uneven road surfaces, coupled with its limited emergency support due to its remote location, made it an inherently dangerous course. Lauda, known for his pragmatic approach to racing, recognized these risks and called for the race to be canceled, citing the perilous conditions. However, despite these concerns, the race proceeded as planned, with teams and organizers ultimately deciding to uphold the event.

Lauda's objections were not driven by fear but by a calculated understanding of the risks involved. His cautious approach to racing prioritized control and safety, a stark contrast to the adrenaline-fueled bravado often associated with the sport. For Lauda, pushing a driver's limits was part of the challenge, but unnecessary risks crossed a line. Despite his reservations, he resolved to race, knowing that any decision to withdraw would jeopardize his championship lead. His dedication to the championship overrode his instincts, setting the stage for a catastrophic accident that would alter his path forever.

The Race Begins: A Dangerous Turn of Events

Race day dawned under cloudy skies, with rain having dampened parts of the track. The inconsistent weather made the Nürburgring even more unpredictable, leaving sections of the track slick and treacherous. As the race began, Lauda and the other drivers faced a circuit that was only partially dry, forcing them to switch from rain tires to slicks early on. This decision, while strategically sound, would prove devastating.

On the second lap, as Lauda approached the challenging Bergwerk corner, his Ferrari suddenly lost traction, veering off the track at high speed. His car collided with a barrier, rebounded onto the track, and burst into flames, trapping Lauda in the cockpit. The impact was severe, tearing off his helmet and exposing him to the flames that engulfed the car. The heat was unbearable, and Lauda struggled to escape, his life hanging by a thread as the inferno intensified around him.

A Heroic Rescue Amidst the Flames

Lauda's life was saved by the heroic actions of fellow drivers Arturo Merzario, Guy Edwards, Brett Lunger, and Harald Ertl, who witnessed the crash and immediately stopped to help. Risking their own lives, they ran to the burning Ferrari, battling against the flames and thick black smoke to pull Lauda from the wreckage. Their courageous efforts prevented an even greater tragedy, but the damage to Lauda was already extensive. His body was severely burned, and he had inhaled toxic fumes that scorched his lungs, leaving him in critical condition.

As Lauda was airlifted to a hospital, the world of Formula 1 was thrown into shock. Drivers, teams, and fans struggled to process the horrific accident, and news of Lauda's injuries spread rapidly. The reality of the dangers inherent in the sport struck a visceral chord, and Lauda's accident became a stark reminder of the thin line between life and death in Formula 1. In the paddock, a somber silence replaced the usual pre-race excitement. For James Hunt, Lauda's closest rival, the tragedy was deeply unsettling, confronting him with the sobering reality that his friend and competitor might never return to racing.

Immediate Impact on Lauda, Hunt, and the Paddock

As Lauda lay in critical condition, hovering between life and death, his absence was felt profoundly within the paddock. Hunt, who had pushed Lauda fiercely all season, was visibly shaken by the accident. The intensity of their rivalry suddenly felt trivial compared to Lauda's fight for survival. Hunt had always admired Lauda's resilience and discipline, but the gravity of the situation brought a new dimension to their relationship. For Hunt, Lauda was more than just an opponent; he was a fellow racer who had risked everything for his love of the sport.

The tragedy also resonated with drivers across the paddock, many of whom had witnessed the dangers of

Nürburgring firsthand. While racing was inherently risky, Lauda's accident served as a stark reminder of just how unforgiving Formula 1 could be. Drivers spoke of their fears, and teams began to question the safety standards that had been in place. Lauda's accident ignited discussions about the need for enhanced safety measures, not only for the cars but also for the circuits on which they raced. The accident revealed a harsh truth: Formula 1, while exhilarating, demanded sacrifices that even the bravest drivers couldn't always anticipate.

The Global Reaction: An Outpouring of Concern

As news of Lauda's condition spread, the world of motorsport, as well as the general public, responded with an outpouring of concern and sympathy. Fans worldwide held vigil, anxiously awaiting updates on Lauda's health. The image of Lauda fighting for his life in a hospital bed became symbolic of the risks drivers faced, and his battle against the odds resonated deeply with fans and non-fans alike. Lauda's accident transcended the boundaries of the sport, highlighting the courage required to be a Formula 1 driver and sparking a broader conversation about safety in motorsport.

In Austria, Lauda's home country, the accident was met with shock and sadness. Lauda was not just a racing champion; he was a national hero, embodying a spirit of resilience and tenacity that Austrians held dear. As his family, friends, and fans rallied around him, the gravity of the accident underscored the extraordinary bravery that had characterized Lauda's career. For those who followed Formula 1, Lauda's accident was a turning point, casting a shadow over the season and intensifying the emotional stakes of his rivalry with Hunt.

Lauda's Fight for Survival

Lauda's injuries were severe. The burns on his face and head were deep, and the damage to his lungs from inhaling toxic fumes made breathing excruciatingly painful. His survival was uncertain, and doctors prepared his family for the worst. Despite his critical condition, Lauda's determination and willpower shone through. Against all odds, he fought to recover, driven by an indomitable spirit and a refusal to let the accident define his career.

In the days following the accident, Lauda's condition fluctuated, but he never gave up. Even in the hospital, he reportedly asked about his condition with a focus on how soon he could return to racing—a testament to his resilience and dedication to the sport. His determination to survive became an inspiration, and as he endured multiple surgeries and intense physical pain, his resolve only grew stronger. Lauda's battle for life became a story of sheer human will, showcasing the strength of character that had defined him as a driver.

The Impact on Formula 1 and Safety Reforms

Lauda's accident was a wake-up call for Formula 1, sparking urgent discussions about safety. While drivers had long accepted the risks of the sport, Lauda's accident highlighted the need for more proactive measures. The call for improved circuit safety standards and car designs grew louder, with drivers and teams alike advocating for change. The "Green Hell" of the Nürburgring became symbolic of the sport's darker side, and the accident accelerated efforts to make Formula 1 safer for future generations.

In the aftermath, Formula 1 authorities began implementing changes, including stricter safety protocols, improved fireproof materials in race suits, and advancements in helmet design. These reforms would gradually transform the sport, reducing the risks that drivers faced while preserving the thrilling essence of Formula 1. Lauda's accident, though tragic, became a catalyst for these critical changes, marking a turning point in the sport's approach to safety.

The Rivalry Transformed: Hunt's Perspective

For James Hunt, Lauda's accident shifted his perspective on their rivalry. The intensity of their competition had been grounded in mutual respect, but the reality of Lauda's brush with death added a new depth to that respect. Hunt, who had often raced with a carefree and fearless attitude, now found himself contemplating the risks they all faced on the track. Though he continued to compete fiercely, he was haunted by the memory of Lauda's accident, and his thoughts often returned to his friend and rival's struggle for recovery.

As Hunt continued to compete, he dedicated several of his races to Lauda, publicly expressing his admiration for the Austrian's bravery and resilience. For Hunt, the accident became a reminder of the humanity that lay beneath their rivalry. He viewed each victory not as a triumph over Lauda but as a testament to the spirit of a man who had risked his life for the sport they both loved.

A Rivalry on Pause – Awaiting Lauda's Return

With Lauda out of the championship for the time being,

Hunt seized the opportunity to close the points gap, winning several races in Lauda's absence. However, even as he celebrated these victories, there was a sense of incompleteness. The rivalry that had defined the season was now hanging in the balance, and Hunt, like many fans, awaited Lauda's return.

The Nürburgring tragedy transformed their rivalry from a mere battle for points into a story of survival, resilience, and mutual respect. Lauda's absence cast a shadow over the season, but his courage in the face of adversity inspired Hunt and the entire racing community. The tragedy at the Nürburgring would forever be remembered as Lauda's defining moment, a testament to the strength of a man who refused to let anything—not even death itself—stand in his way.

As the season continued, Formula 1 waited with bated breath to see if Lauda would return. His comeback, should it happen, would be nothing short of miraculous, adding an unforgettable chapter to a rivalry that had already transcended the boundaries of the sport.

Chapter 7: The Fighter Returns – Lauda's Remarkable Comeback

The world of Formula 1 had barely begun to recover from the shock of Niki Lauda's near-fatal accident at the Nürburgring when rumors started circulating about his return. Only six weeks had passed since Lauda had fought for his life in a hospital bed, and yet, despite his still-raw injuries and the searing pain, he was preparing to race again. His comeback at the Italian Grand Prix at Monza became one of the most inspiring stories in motorsport history, marking a defining moment not only for Lauda but also for his rivalry with James Hunt. In that race, the tone of their competition shifted profoundly, as mutual respect tempered their fierce rivalry.

An Unbelievable Recovery

After the horrifying accident, Lauda's condition had been touch and go. The burns that scarred his face, head, and scalp were severe, and the damage to his lungs from inhaling toxic fumes had left doctors doubtful of his survival, let alone his ability to race again. Lauda, however, possessed a fierce will to survive. Even in his most critical moments, his thoughts were reportedly on racing, and as soon as he was able, he began pushing himself to recover, defying medical advice and his own physical limitations. His sheer determination astounded his medical team, who saw a man whose mind was set firmly on returning to Formula 1.

The scars were visible, and the pain was ever-present, but Lauda remained undeterred. For Lauda, the choice to return to the track was about reclaiming his life and control over his career. He refused to let the accident define him or end his journey in the sport he loved. But it wasn't just physical endurance that made Lauda's comeback possible; it was his unbreakable spirit. With each painful step of his recovery, Lauda proved that his courage and resilience ran deeper than any fear of physical suffering or death.

Arriving at Monza – The Paddock's Reaction

Lauda's announcement that he would compete at the Italian Grand Prix at Monza, only six weeks after the accident, sent shockwaves through the paddock. Fellow drivers, team members, and fans couldn't believe it; some worried about his health, while others were inspired by his bravery. Formula 1 had seen remarkable displays of courage before, but Lauda's comeback was unprecedented. His very presence at Monza was a statement of defiance against fear and adversity, a message that he would not be defined by his injury.

As Lauda arrived at the track, the world saw a changed man. His face, still bandaged and scarred, bore the physical marks of his ordeal. He wore a specially designed helmet that accommodated his bandages, and his movements were visibly cautious, each step a reminder of his painful journey. Yet beneath the surface was the same fierce determination that had driven him to become World Champion. Lauda's return wasn't just about competing; it was about reclaiming his identity as a driver and as a champion.

For James Hunt, seeing Lauda at Monza was both humbling and deeply moving. Hunt had been winning races in Lauda's absence, closing the gap in the championship standings, but Lauda's return reminded him that the battle between them was far more than just a competition for points. Hunt recognized the extraordinary courage it took for Lauda to race, and he understood that their rivalry was now something deeper, grounded in mutual respect as much as in their shared desire to win.

The Italian Grand Prix – A Courageous Performance

The race at Monza was set to be one of the most challenging of Lauda's career. His body was still recovering, and he was in constant pain. Yet, when the lights went out, Lauda drove with a resilience and focus that defied his physical limitations. He qualified fourth, an incredible feat considering his condition, and as the race began, he maintained a pace that stunned his competitors. Each lap was a testament to Lauda's unwavering will, as he fought through the pain to keep up with the leaders.

Lauda finished the race in fourth place, earning valuable championship points and proving that he was still a force to be reckoned with. His performance was met with applause from fans, competitors, and team members alike. For Lauda, the race at Monza was about more than points; it was a victory over fear, a defiance of the physical and emotional scars left by his accident. The crowd, initially unsure of what to expect, erupted in admiration for Lauda's courage. He had defied all odds, demonstrating that his spirit was as indomitable as ever.

For Hunt, who finished the race in third, Lauda's return added a new dimension to their rivalry. No longer was it simply about competing for the championship; it was about the resilience and dedication that defined both men. Hunt, who had raced fiercely in Lauda's absence, now saw his competitor in a new light, acknowledging that the fight for the title was now a battle between two equally driven men, each willing to sacrifice everything to prove their worth.

A Rivalry Recast in Respect

Lauda's remarkable comeback at Monza recast the rivalry between him and Hunt. While they remained fierce competitors, their relationship evolved into one of mutual respect, grounded in the shared understanding of what it meant to risk everything for the sport they loved. Hunt's admiration for Lauda's resilience was clear, and Lauda, in turn, respected Hunt's competitive spirit and the manner in which he had seized his opportunities during Lauda's absence. The paddock, sensing the shift, witnessed a rivalry transformed; their competition was now less about points and more about the courage and resolve required to be a champion.

This newfound respect didn't dampen the intensity of their rivalry; if anything, it amplified it. Both men recognized that the championship would be decided by who was willing to push themselves the furthest, to endure the most. For Hunt, the sight of Lauda on the track served as a reminder of the resilience he would need to claim the title. For Lauda, every lap he completed was a statement to Hunt, and the world, that he was not merely surviving—he was still fighting to win.

The Championship Race Tightens

With Lauda's return, the 1976 championship battle was reignited. The points gap between Hunt and Lauda was narrow, and each race became a high-stakes showdown. Both drivers knew that any mistake could cost them the title, and their rivalry took on a new intensity as they pushed themselves to the absolute limit. Fans followed every turn with heightened anticipation, witnessing a battle that was as much about survival and resilience as it was about racing skill.

As the season continued, each race became a contest of willpower and endurance. Hunt and Lauda's mutual respect drove them to push harder, knowing that they were not just racing for themselves but also competing against an opponent they genuinely admired. The tone of the championship had shifted from one of animosity to one of shared purpose, as both men sought to prove not only who was the better driver but also who possessed the greater strength of character.

The Legacy of Lauda's Comeback

Lauda's comeback at Monza left a lasting impact on Formula 1 and on Hunt. It marked a turning point in the way drivers approached the sport, highlighting the resilience and courage required to overcome the dangers of racing. Lauda's return had proven that a driver's strength was not just physical but also deeply rooted in mental and emotional resilience. His determination inspired fans and competitors alike, serving as a reminder that champions are defined not only by their victories but by their ability to confront adversity head-on.

For Hunt, Lauda's comeback reinforced the essence of their rivalry. Their battle for the championship became a contest not just of skill but of willpower, resilience, and the courage to confront one's fears. Hunt, who had once seen Lauda as a disciplined yet distant competitor, now understood the depth of his rival's character, recognizing that they were bound by the same relentless pursuit of excellence.

Setting the Stage for the Final Showdown

With the championship battle back in full force, Hunt and Lauda entered the remaining races with a renewed sense of purpose. Each race was a test of endurance and mental fortitude, with the two men competing not only to win but also to honor the spirit of their rivalry. Lauda's comeback had redefined their competition, transforming it into a story of resilience, respect, and shared ambition.

As the season neared its conclusion, Hunt and Lauda both knew that the championship would come down to the final race. The tension was palpable, with fans around the world captivated by the story of two men who had given everything they had—physically, mentally, and emotionally—for a chance at victory. The final race at Fuji would decide the 1976 World Champion, but for Hunt and Lauda, it was about more than just a title. It was about proving, to themselves and to each other, who truly embodied the spirit of a champion.

Lauda's return at Monza marked the beginning of a new chapter in their rivalry—a chapter that would culminate in one of the most dramatic finales in Formula 1 history. The stage was set for a final showdown, where courage, respect, and the unbreakable will to win would define not only the race but also the legacies of two of the sport's greatest competitors.

Chapter 8: Fuji Showdown – The Battle in the Rain

The 1976 Formula 1 season had been nothing short of a saga, with Niki Lauda and James Hunt fighting tooth and nail across circuits worldwide, pushing themselves to the limits of endurance, resilience, and skill. After months of heated competition, both men arrived at the final race of the season, the Japanese Grand Prix at Fuji, with everything to play for. For Hunt, a chance at his first world title; for Lauda, the defense of a title he had fought for, almost dying in the process. Their rivalry, fueled by mutual respect and an unbreakable will, was now poised to reach its dramatic climax.

The Setting – Fuji Speedway

The Fuji Speedway, located near Mount Fuji's picturesque, snow-capped slopes, presented a daunting 2.7-mile circuit, known for its long straights and technical corners. It was an unforgiving track that demanded focus and courage. As the weekend unfolded, so did a tension that simmered through the paddock, with everyone aware that the World Championship would be decided on this track, between two men whose rivalry had captivated the world.

As the teams arrived at Fuji, the air was thick with anticipation. Formula 1 fans, drawn by the season's dramatic narrative, turned out in record numbers, braving the cold October air. They sensed that the showdown between Hunt and Lauda would be one for the history books, a contest that transcended the sport. However, none could have anticipated the drama that the weather would add to the day.

The Downpour – A Test of Nerves

Race day dawned under ominous skies. Dark clouds hung heavily over Fuji, and by the time the drivers assembled on the grid, a relentless rain had set in, soaking the track and obscuring visibility. Streams of water poured across the tarmac, creating treacherous conditions and raising doubts about the safety of the race. The rain was relentless, transforming Fuji into a slick, dangerous arena where even the smallest mistake could be catastrophic.

As drivers, team members, and officials evaluated the conditions, the question arose: should the race be delayed or even canceled? The visibility was so poor that drivers could barely see the end of their own cars, let alone the vehicle in front of them. Many voiced concerns, but the decision ultimately came down to the event organizers. Despite the potential risks, and with pressure from broadcasters and fans eager to see the showdown, the race was set to go ahead. Hunt and Lauda, like the other drivers, had no choice but to line up on the rain-soaked grid, nerves steeled for what would be one of the most grueling races of their lives.

The Start – Hunt's Determination, Lauda's Caution

When the lights went out, Hunt surged forward, charging into the race with a characteristic fearlessness that bordered on recklessness. The championship was within his reach, and he was willing to risk everything to seize it. Lauda, by contrast, was cautious, acutely aware of the treacherous conditions. Just months before, he had suffered a life-altering crash, and though his courage was undeniable, he had come to value his life and well-being above all else.

The cars thundered through the downpour, wheels spinning and visibility close to zero. Hunt's McLaren slipped and skidded, but he muscled it through each corner, narrowly avoiding spins that might have ended his championship bid right there. The spray from the cars created a foggy wall of water, obscuring the view for those behind him, and the risks multiplied with each lap. Hunt's aggressive style, so well-suited to dry tracks, was a high-stakes gamble in these conditions, but he powered forward, determined to push through.

Lauda, meanwhile, wrestled with a different set of calculations. As the laps went on, his caution turned into concern, and he began questioning whether the risks were worth it. The rain was relentless, and each corner felt like a tightrope walk over a chasm. Every instinct told him that pushing further could mean more than just a lost race—it could mean losing his life.

Lauda's Decision – The Choice to Withdraw

On the third lap, Lauda made a decision that stunned the world. With the championship still on the line, he pulled his Ferrari into the pits, choosing to withdraw from the race. It was a decision born not out of fear, but of deep conviction and wisdom—a choice to prioritize life over glory, a rare display of vulnerability and strength in a sport that demanded a relentless drive to the finish. Lauda's withdrawal was a statement that he would not let racing define the limits of his life, that he valued himself as more than just a competitor.

As Lauda climbed out of his car, the world watched in shock. Many couldn't understand how a reigning champion, a driver who had fought back from the brink of death, could make such a decision. But for Lauda, this was the ultimate act of courage. He knew he could continue and perhaps fight for the championship, but he refused to compromise his principles and safety. Formula 1 had nearly claimed his life once; he wasn't about to let it do so again.

Hunt, still racing, had no idea of Lauda's decision. Focused solely on the track, he pushed through the rain, his determination unwavering. To him, every lap was a battle for survival and victory. He knew that Lauda was his primary competitor, and without realizing Lauda was no longer in the race, he continued to fight as though his rival were right behind him.

The Battle on the Track – Hunt's Struggle

Hunt's path to the finish was anything but easy. The treacherous conditions demanded every ounce of his skill and concentration. He was in a fierce battle with Mario Andretti and Jody Scheckter, who were both vying for the lead, making Hunt's path to the championship even more challenging. His McLaren skidded, hydroplaned, and seemed on the verge of slipping off the track with each corner, but Hunt's fierce drive kept him going, refusing to let the rain or the mounting exhaustion break his resolve.

With Lauda out of the race, Hunt's odds of winning the championship had improved, but he still needed to finish high in the standings. Every lap was a fight against time, as he calculated his position and strategized his moves with a singular goal in mind: securing enough points to take the championship. The rain continued to pound down, visibility remained low, and Hunt's hands clenched the wheel as he pushed his car through the waterlogged track.

In the final laps, Hunt's position slipped slightly, and it seemed as though his chance at the championship might be slipping away with it. But with remarkable grit, he clawed his way back, fighting for every inch, every fraction of a second, as he edged toward the finish line. The final laps were a blur of rain, speed, and sheer determination as Hunt pushed his car beyond its limits, determined to seize the title.

Crossing the Finish Line – A Confusing Victory

As Hunt crossed the finish line, there was a moment of confusion. Unsure of his exact position due to the chaotic conditions and the intense battle on the track, Hunt didn't initially realize that he had secured the points needed to clinch the championship. Exhausted, soaked, and emotionally drained, he climbed out of his car with no idea that he had achieved his lifelong dream.

It was only after a few tense minutes that the news reached him: James Hunt was the 1976 World Champion. The realization hit him with a mix of elation, relief, and disbelief. After months of battling Lauda, through the highs of victory and the depths of frustration, Hunt had finally achieved the pinnacle of Formula 1. The triumph, though sweet, was also tempered by the emotional toll of the season and the knowledge that he had won in extraordinary, almost surreal circumstances.

Lauda and Hunt – A Rivals' Respect

For Lauda, watching from the sidelines, Hunt's victory was a bittersweet moment. Though he had relinquished the title, he felt no bitterness. Lauda knew he had made the right choice for himself, and he admired Hunt's grit in securing the championship. Their rivalry, defined by fierce competition, was now over, but it left a legacy built on mutual respect, resilience, and an unspoken bond. Lauda, who had risked everything for Formula 1, now understood the value of limits, while Hunt, who had fought relentlessly for the title, had won not only a championship but also the admiration of his greatest competitor.

Their rivalry had reached its conclusion on a rain-soaked track in Japan, under circumstances neither could have predicted. Yet, in many ways, the Fuji showdown was the perfect ending to their season. It was a race that tested their courage, their endurance, and their sense of self. For Hunt, it was a moment of triumph; for Lauda, a moment of clarity.

The Legacy of Fuji – A Defining Moment in Formula 1

The Fuji race became one of the most iconic moments in Formula 1 history, symbolizing both the glory and the risk inherent in the sport. Hunt's victory and Lauda's withdrawal spoke volumes about the human spirit, about the line between courage and recklessness, and about the strength it took to make decisions that defied the expectations of the world.

For fans, Fuji was more than just a race; it was the culmination of a season defined by passion, resilience, and respect. Hunt and Lauda had pushed each other to the very limits, and in doing so, had raised Formula 1 to new heights. They had shown that a rivalry could be fierce without being toxic, that competition could coexist with admiration, and that sometimes, the greatest act of courage was knowing when to step away.

The 1976 season would be remembered not only for the championship Hunt won but for the lessons that Lauda's comeback and decision at Fuji imparted. Their battle in the rain was more than just a race; it was a testament to the spirit of two men who, despite their differences, had found a connection on the track that went beyond winning or losing. And as the rain subsided and the engines cooled, the world of Formula 1 was forever changed by the legacy of the Fuji showdown.

Chapter 9: Beyond the Podium – The Legacy of Hunt vs. Lauda

The 1976 Formula 1 season left a mark that went far beyond the circuits and the final standings. It was a year defined by grit, rivalry, respect, and the human spirit's resilience, embodied by James Hunt and Niki Lauda. Their championship battle, punctuated by moments of intense competition, personal sacrifice, and mutual respect, resonated with millions of fans worldwide. In the years that followed, their story became legendary, shaping Formula 1's culture, inspiring a new generation of fans, and establishing a benchmark for sportsmanship that continues to echo across the sport.

The Duality of Hunt and Lauda

James Hunt and Niki Lauda couldn't have been more different, yet it was precisely this contrast that made their rivalry so captivating. Hunt, the charismatic British playboy, brought an element of passion and spontaneity to Formula 1 that endeared him to fans. His freewheeling personality, devil-may-care attitude, and unfiltered honesty made him a larger-than-life character, embodying the rebellious spirit of the 1970s. Hunt's approach to racing was visceral and instinctual, and his style struck a chord with those who believed racing was about more than just technical mastery—it was about heart, bravery, and the thrill of the moment.

Lauda, in contrast, was the epitome of precision and professionalism. Disciplined, methodical, and intensely focused, Lauda represented a new breed of driver who brought scientific rigor to the sport. His approach was cool and calculated, often misunderstood as cold or aloof, but it was rooted in an unwavering commitment to excellence. Lauda's technical understanding of his car, his close relationships with his engineers, and his methodical race preparation set a new standard in Formula 1, influencing countless drivers who would follow.

Together, they represented two sides of the same coin—passion and precision, chaos and control. This duality captured fans' imaginations, transforming their rivalry into a narrative that transcended the sport. Hunt and Lauda weren't just racing each other; they were racing against their own ideals, philosophies, and identities. The 1976 season crystallized this dichotomy, and their struggle for the title became a story of competing values, making Formula 1 about more than just speed and skill; it was about the essence of what it meant to be a champion.

The Impact on Formula 1 Culture

The rivalry between Hunt and Lauda redefined what it meant to be a Formula 1 driver. Their contrasting personalities and racing styles illustrated that there was no one "right" way to be a champion. For some drivers, it was about pushing limits and taking risks, as Hunt did. For others, it was about calculated control and consistency, as Lauda exemplified. This realization opened the sport to a broader range of personalities and driving styles, celebrating diversity in approaches rather than adhering to a singular image of what a Formula 1 driver should be.

The story of Hunt and Lauda introduced a spirit of sportsmanship that still reverberates through the paddock today. Despite their fierce rivalry, they never let competition breed animosity. Hunt openly admired Lauda's courage, especially after Lauda's miraculous return, while Lauda respected Hunt's relentless determination and passion for the sport. Their relationship set a powerful example, showing that rivals could respect one another, even in the midst of intense competition. It established a standard for what it meant to compete in Formula 1—a standard that celebrated victory but also valued resilience, integrity, and camaraderie.

A New Generation of Fans

The 1976 season brought a flood of new fans to Formula 1, many drawn by the dramatic, almost cinematic, storyline between Hunt and Lauda. The media coverage of their rivalry, Lauda's fiery accident and subsequent comeback, and Hunt's ultimate triumph in the rain-soaked Fuji showdown captured the public's imagination like never before. Formula 1 became more than just a motorsport; it was a spectacle, a dramatic saga played out on some of the world's most dangerous tracks.

This narrative helped solidify Formula 1's place in popular culture, attracting fans who might not have otherwise paid attention to the sport. Hunt and Lauda became symbols of courage and resilience, inspiring those who saw racing as more than just cars and tracks. The 1976 season's storytelling appeal laid the groundwork for the way Formula 1 would be covered in the future, emphasizing personal narratives, rivalries, and the human drama that unfolded on the track. Hunt and Lauda's story showed that Formula 1 could tell stories that reached beyond motorsport, touching on universal themes of fear, ambition, triumph, and survival.

Legacy of Resilience and Recovery

Niki Lauda's recovery and comeback after his near-fatal crash at the Nürburgring set a precedent for resilience in Formula 1. His determination to return to racing so soon after suffering life-threatening injuries redefined what it meant to be a Formula 1 driver. Lauda's strength, courage, and refusal to let fear control his life inspired generations of drivers and fans alike. His comeback made Formula 1 safer, accelerating the push for better protective equipment, fireproof suits, improved helmets, and overall track safety.

For many drivers, Lauda's comeback became a benchmark for courage. Formula 1 was a sport of high risks, but Lauda's decision to return showed that strength was not only physical but deeply rooted in mental resolve. Drivers began to see themselves not just as racers but as resilient athletes who could overcome adversity. Lauda's legacy of resilience became a fundamental part of Formula 1's ethos, setting a standard that continues to shape the sport today.

Hunt's Influence on the Spirit of Formula 1

James Hunt, on the other hand, became an icon of passion, a reminder that Formula 1 was a sport where risks were taken and rules occasionally bent in pursuit of glory. Hunt's fearless, all-or-nothing style inspired drivers who wanted to race with their hearts as much as with their heads. His charismatic personality and candid approach to life made him a relatable hero for fans who valued freedom and spontaneity. Hunt showed that Formula 1 wasn't just about careful planning; it was also about seizing the moment, about the raw joy of racing.

Hunt's influence can be seen in drivers who embrace the thrill of the race and aren't afraid to be themselves. His unapologetic honesty and charisma left an indelible mark on Formula 1's spirit, reminding the sport that personality mattered as much as performance. For fans, Hunt was a reminder of the pure, unrestrained love of racing—a quality that many still look for in their heroes on the grid today.

A Legacy of Mutual Respect and Sportsmanship

Perhaps the most enduring legacy of the Hunt-Lauda rivalry is the sportsmanship and respect they showed each other. Despite their differences and the intensity of their rivalry, Hunt and Lauda never allowed competition to erode their admiration for one another. They understood the sacrifices they both made and respected each other's courage in the face of life-threatening risks.

Their rivalry showed that sportsmanship and mutual respect were not just ideals but achievable standards, even at the highest level of competition.

Their friendship after the 1976 season became a poignant aspect of their story. Lauda and Hunt remained close until Hunt's untimely death in 1993, often reminiscing about their days on the track and the unforgettable season they shared. For Lauda, Hunt was more than just a competitor; he was a friend who had shaped his life and career. For Hunt, Lauda was a man of unparalleled discipline and strength, someone he respected deeply even as he fought him for the championship. Their bond became a reminder to fans and future drivers alike that true rivalry can coexist with true respect.

Defining the Essence of Formula 1

The Hunt-Lauda rivalry captured the essence of Formula 1 as a sport that combined technical brilliance with human courage. Their battle in 1976 showed that Formula 1 was more than just about who was fastest; it was a sport where drivers confronted their fears, battled with themselves, and often faced the ultimate risk in pursuit of victory. Hunt and Lauda's rivalry illustrated that racing was not only a physical contest but a psychological one, where character, courage, and values were just as important as skill and speed.

For Formula 1, the 1976 season became a benchmark for what the sport could represent. It wasn't just about winning trophies; it was about pushing boundaries, embracing challenges, and embodying the spirit of competition at its most intense. Hunt and Lauda's rivalry set a standard for sportsmanship and resilience that drivers still look up to, inspiring them to honor both the thrill and the inherent risks of Formula 1 with respect and reverence.

A Story That Endures

The legacy of Hunt and Lauda lives on, both in the annals of Formula 1 history and in the hearts of fans who remember their story. Their rivalry, filled with drama, respect, and resilience, continues to inspire, reminding the world that sport is a stage where life's greatest values—passion, respect, courage, and determination—play out. Their story has been retold in books, documentaries, and movies, reaching new generations who see in Hunt and Lauda not only champions but symbols of an era when Formula 1 was as much about personality as it was about performance.

Beyond the podium, beyond the points, and beyond the years, the legacy of Hunt and Lauda endures as a tale of two men who, despite their differences, shared a bond forged in the fires of competition. They remain icons of Formula 1, forever remembered as rivals who transformed each other and the sport itself, leaving behind a legacy that will continue to inspire for generations to come.

Chapter 10: Legends Forged in Fire – The Influence of a Rivalry

The 1976 season between James Hunt and Niki Lauda has become one of the most enduring stories in Formula 1 history, a rivalry that transcended the sport itself and entered the realm of legend. Decades later, the influence of their battle is still felt, inspiring drivers, shaping modern Formula 1, and reinforcing the values of sportsmanship, resilience, and mutual respect. Hunt and Lauda's story was forged in the intensity of their rivalry, and its lasting impact speaks to the power of their characters and the lessons they left behind.

A Rivalry Remembered

For Formula 1 fans, the Hunt-Lauda rivalry remains a story of pure drama, of two men who approached racing with diametrically opposed philosophies yet shared a mutual respect that was rare among rivals. Hunt, the charismatic, free-spirited Brit with a taste for risk and flair, and Lauda, the disciplined, calculated Austrian who valued precision over bravado, represented the two sides of what racing could be. They both brought out the best in each other, and their contest became a narrative that showcased Formula 1 as a sport where courage, intellect, and emotion collided.

Their rivalry is still celebrated today in films, documentaries, and books that capture the essence of their competition. The 2013 film *Rush*, directed by Ron Howard, brought their story to a new generation of fans, reigniting interest in their lives, their contrasting styles, and the unforgettable 1976 season. The film, like the season itself, emphasized that rivalry in sport could be fierce yet honorable, a bond that connected Hunt and Lauda long after their days on the track were over. This story of rivalry and respect continues to resonate with audiences, reminding fans that even the most intense competition can exist alongside admiration and understanding.

The Influence on Modern Formula 1

Formula 1 has changed dramatically since 1976, evolving into a sport where technology and data analysis play as critical a role as driver skill. Yet the Hunt-Lauda rivalry remains a benchmark for what Formula 1 aspires to be: a competition not only of machines but of personalities, where drivers' individual philosophies and strengths bring depth and dimension to the race. Their legacy has encouraged modern drivers to bring their unique personalities to the grid, reminding them that Formula 1 is not just about winning but about how you win.

In an era when the sport has become more globalized, the Hunt-Lauda rivalry provides a timeless model for sportsmanship. Modern drivers like Lewis Hamilton, Sebastian Vettel, and Fernando Alonso have spoken of Hunt and Lauda with reverence, recognizing the influence of their rivalry on how they approach competition. Today's Formula 1 drivers are highly skilled athletes who understand the value of mental resilience and dedication, qualities exemplified by Lauda. At the same time, they embrace Hunt's philosophy of racing with passion and personal expression, recognizing that fans are drawn to the human side of racing as much as to the technical prowess.

The values embodied by Hunt and Lauda have set a standard for respectful rivalry in Formula 1, encouraging drivers to view their competitors as partners in a shared journey. This legacy has helped build a culture in modern Formula 1 where competition is fierce but sportsmanship is paramount, where rivals shake hands and celebrate each other's successes. It is a reminder that Formula 1, while intensely competitive, is also a community of people who understand the sacrifices and risks each driver faces on the track.

A Catalyst for Safety Improvements

Lauda's accident at the Nürburgring was a watershed moment that catalyzed significant safety reforms in Formula 1. His fiery crash underscored the dangers of racing and highlighted the need for stronger safety protocols to protect drivers. The accident became a turning point for the sport, accelerating the development of measures that would make Formula 1 safer for generations to come. As Lauda himself later reflected, his accident was painful, but it served a purpose by making Formula 1 recognize the urgent need for change.

In the years following the 1976 season, Formula 1 implemented a series of safety improvements that transformed the sport. The introduction of fire-resistant race suits, improved helmet designs, and better medical facilities at tracks were direct responses to Lauda's accident. The FIA (Fédération Internationale de l'Automobile) also began enforcing stricter safety standards for circuits, reducing the number of dangerous tracks and upgrading crash barriers to prevent similar incidents.

Over the years, this commitment to safety has only intensified, with innovations like the HANS (Head and

Neck Support) device and, more recently, the Halo cockpit protection system. These advancements are a tribute to Lauda's legacy, to the sacrifices he made, and to his unyielding demand that drivers be protected to the fullest extent possible. The sport's dedication to safety stands as a testament to the lessons learned from the Hunt-Lauda rivalry, underscoring Formula 1's responsibility to safeguard its drivers.

Legacies That Endure

James Hunt and Niki Lauda may no longer be with us—Hunt passed away in 1993 at the age of 45, and Lauda in 2019 at 70—but their legacies continue to inspire Formula 1 and beyond. Hunt, the charismatic playboy who lived for the thrill, remains an icon of passion and individuality. His approach to life and racing, unapologetically bold and daring, inspires fans who believe in the joy of living each moment to its fullest. Hunt's charisma and willingness to be himself in an often-staid sport encouraged Formula 1 to embrace a broader range of personalities, making space for drivers who wanted to express themselves beyond the race track.

Lauda's legacy, meanwhile, is one of resilience, courage, and a commitment to excellence. His disciplined approach to racing, his technical acumen, and his relentless pursuit of improvement have set a standard that continues to influence Formula 1 drivers today. Lauda's life was a testament to the power of grit and determination, and his remarkable comeback at Monza has become a story of survival and perseverance that resonates deeply in the racing community. For Formula 1, Lauda remains a symbol of the mental toughness required to succeed, a reminder that champions are forged not only by their victories but by how they respond to adversity.

Together, Hunt and Lauda's legacies represent the duality of Formula 1 as a sport that celebrates both the thrill of competition and the depth of personal character. Their rivalry, intense and unforgettable, stands as a testament to the spirit of racing, where the desire to win is balanced by respect, resilience, and the courage to stay true to oneself. Their influence has woven a sense of history and humanity into Formula 1, giving fans more than just records to admire, but personalities, values, and stories that transcend the sport.

The Timelessness of Their Story

The story of Hunt and Lauda is timeless, resonating with fans who see in them the embodiment of Formula 1's core values. Their rivalry, forged in the fires of competition, has become a symbol of what makes the sport compelling—the unpredictable drama, the fierce passion, and the respect between competitors who understand the risks they take. Their battle in 1976 remains one of the most celebrated seasons in Formula 1 history, not only for the spectacle it provided but for the humanity it revealed.

Their story has inspired not only drivers and fans but also a broader audience who see in Hunt and Lauda's rivalry a powerful narrative of resilience, friendship, and the courage to follow one's path. It has shown that sportsmanship can elevate competition into something profound, that rivalries can create bonds that endure beyond the finish line. For those who love Formula 1, the legacy of Hunt and Lauda is a reminder that racing is more than just a sport—it is a stage for the exploration of the human spirit, a place where courage, rivalry, and respect intertwine.

A Legacy Forged in Fire

The rivalry between James Hunt and Niki Lauda, forged in the crucible of one unforgettable season, continues to shape the way Formula 1 views itself and the values it upholds. Their story has left an indelible mark on the sport, a legacy that endures in every driver who straps into a cockpit, every fan who cheers from the stands, and every moment of triumph and tragedy on the track. They were, and remain, legends in Formula 1—a testament to the idea that the true essence of sport lies not just in victory, but in the character, resilience, and respect that define those who compete.

In the end, Hunt and Lauda taught us that rivalry could be a source of inspiration, that courage could coexist with caution, and that victory was sweetest when earned with integrity. Their influence extends beyond the podium and beyond the years, capturing the hearts of those who admire not only champions but also the character and spirit that champions bring to their sport. In Formula 1, Hunt and Lauda will forever be remembered as rivals who pushed each other, respected each other, and, in doing so, left a legacy that will live on as long as the engines roar and the lights go out.

About the Author

Etienne Psaila, an accomplished author with over two decades of experience, has mastered the art of weaving words across various genres. His journey in the literary world has been marked by a diverse array of publications, demonstrating not only his versatility but also his deep understanding of different thematic landscapes. However, it's in the realm of automotive literature that Etienne truly combines his passions, seamlessly blending his enthusiasm for cars with his innate storytelling abilities.

Specializing in automotive and motorcycle books, Etienne brings to life the world of automobiles through his eloquent prose and an array of stunning, high-quality color photographs. His works are a tribute to the industry, capturing its evolution, technological advancements, and the sheer beauty of vehicles in a manner that is both informative and visually captivating.

A proud alumnus of the University of Malta, Etienne's academic background lays a solid foundation for his meticulous research and factual accuracy. His education has not only enriched his writing but has also fueled his career as a dedicated teacher. In the classroom, just as in his writing, Etienne strives to inspire, inform, and ignite a passion for learning.

As a teacher, Etienne harnesses his experience in writing to engage and educate, bringing the same level of dedication and excellence to his students as he does to his readers. His dual role as an educator and author makes him uniquely positioned to understand and convey complex concepts with clarity and ease, whether in the classroom or through the pages of his books.

Through his literary works, Etienne Psaila continues to leave an indelible mark on the world of automotive literature, captivating car enthusiasts and readers alike with his insightful perspectives and compelling narratives.

Visit www.etiennepsaila.com for more.